SAE EDGE™
RESEARCH REPORT

Unsettled Issues in Balancing Virtual, Closed-Course, and Public-Road Testing of Automated Driving Systems

Sven Beiker, PhD
Silicon Valley Mobility

EDGE DEVELOPMENT TEAM

Atul Acharya, *AAA/A3Ventures*
Michał Antkiewicz, *University of Waterloo*
David Barry, *Multek Technologies*
Thomas Bock, *Samsung*
Krzysztof Czarnecki, *University of Waterloo*
Eric Paul Dennis, PE, *Center for Automotive Research*
Cameron Gieda, *AutonomouStuff*
Peter-Nicholas Gronerth, *fka Silicon Valley*
Qiang Hong, PhD, *Center for Automotive Research*
Stefan Merkl, *TÜV SÜD America*
Chad Partridge, *Metamoto*
Robert Seidl, *Motus Ventures*
John Suh, *Hyundai CRADLE*
John Tintinalli, *SAE International*
Paul Wells, *AAA/A3Ventures*

SAE INTERNATIONAL®
Warrendale, Pennsylvania, USA

About the Publisher

SAE International® is a global association of more than 128,000 engineers and related technical experts in the aerospace, automotive, and commercial-vehicle industries. Our core competencies are lifelong learning and voluntary consensus standards development. Visit sae.org

SAE EDGE™ Research Report Disclaimer

SAE EDGE™ Research Reports focus on topics that are dynamic, in which knowledge is incomplete, and which have yet to be standardized. They represent the collective wisdom of a group of experts and serve as a practical guide to the reader in understanding unsettled subject matter. They are not meant to provide a recommended practice or protocol. The experts engaged have contributed their own thoughts and points of view and these are not the positions of the institutions or businesses with which they are affiliated. A professional writer has collectivized their input; there is no one contributor's perspective being advanced, but rather that of a community of practitioners. SAE EDGE™ Research Reports are the property of SAE International and SAE alone is responsible for their content.

About This Publication

SAE EDGE™ Research Reports provide state-of-the-art and state-of-the-industry examinations of the most significant topics in mobility engineering. SAE EDGE™ contributors are experts from research, academia, and industry who have come together to explore and define the most critical advancements, challenges, and future direction in areas such as vehicle automation, unmanned aircraft, Internet of Things, and connectivity, cybersecurity, advanced propulsion, and advanced manufacturing.

Related Resources

SAE MOBILUS® Automated & Connected Knowledge Hub
http://sae.org/Mobilus/Automated

SAE Team

Frank Menchaca, Chief Product Officer
Michael Thompson, Director, Standards, Information and Research Publications
Monica Nogueira, Acquisitions Director
Jill Leonard, Product Manager
William Kucinski, Managing Technical Editor

Printed in USA

ISSN 2640-3536
e-ISSN 2640-3544
ISBN 978-1-4686-0125-1

To purchase bulk quantities, please contact: SAE Customer Service

E-mail: CustomerService@sae.org
Phone: 877-606-7323 (*inside USA and Canada*)
724-776-4970 (*outside USA*)
Fax: 724-776-0790

Visit the SAE International Bookstore at books.sae.org

About the Editor

Sven Beiker, PhD, is the Founder and Managing Director of Silicon Valley Mobility, a mobility consulting and advisory firm that specializes in technical diligence, product roadmaps, and business models for mobility topics. The engagements span startups, investors, and corporations in the mobility and adjacent industries. In addition, Dr. Beiker is a Lecturer at the Graduate School of Business at Stanford University where he instructs students on strategies for startups and corporations in the field of automated, connected, electrified, and shared mobility.

With his well over 20 years of experience gained during his tenure at McKinsey & Company, Stanford University, and the BMW Group, Dr. Beiker is dedicated to the future of the automobile and personal mobility. His mission is to improve sustainability, safety, efficiency, and convenience in how consumers use automobiles. He combines perspectives from technology, business, policy, and human factors.

Dr. Beiker also serves on advisory boards of several startups in the mobility space, as an advisor to the German American Chamber of Commerce in San Francisco, and as an advisor/co-editor to the Lecture Notes in Mobility of Springer Science+Business Media. He is a continuing contributor to the SAE EDGE™ Research Report series.

contents

Unsettled Issues in Balancing Virtual, Closed-Course, and Public-Road Testing of Automated Driving Systems

Abstract

This SAE EDGE™ Research Report identifies key unsettled issues of interest to the automotive industry regarding the challenges of determining the optimal balance for testing automated driving systems (ADS). Three main issues are outlined that merit immediate interest:

First, determining what kind of testing an ADS needs before it is ready to go on the road.

Second, the current, optimal, and realistic balance of simulation testing and real-world testing.

Third, the challenges of sharing data in the industry.

SAE EDGE™ Research Reports are preliminary investigations of new technologies. The three technical issues identified in this report should be discussed in greater depth with the aims of, first, clarifying the scope of the industry-wide alignment needed; second, prioritizing the issues requiring resolution; and, third, creating a plan to generate the necessary frameworks, practices, and protocols.

NOTE: SAE EDGE™ Research Reports are intended to identify and illuminate key issues in emerging, but still unsettled, technologies of interest to the mobility industry. The goal of SAE EDGE™ Research Reports is to stimulate discussion and work in the hope of promoting and speeding resolution of identified issues. SAE EDGE™ Research Reports are not intended to resolve the issues they identify or close any topic to further scrutiny.

SVEN BEIKER, PhD
Silicon Valley Mobility

Edge Development Team

Atul Acharya, *AAA/A3Ventures*
Michał Antkiewicz, *University of Waterloo*
David Barry, *Multek Technologies*
Thomas Bock, *Samsung*
Krzysztof Czarnecki, *University of Waterloo*
Eric Paul Dennis, PE, *Center for Automotive Research*
Cameron Gieda, *AutonomouStuff*
Peter-Nicholas Gronerth, *fka Silicon Valley*
Qiang Hong, PhD, *Center for Automotive Research*
Stefan Merkl, *TÜV SÜD America*
Chad Partridge, *Metamoto*
Robert Seidl, *Motus Ventures*
John Suh, *Hyundai CRADLE*
John Tintinalli, *SAE International*
Paul Wells, *AAA/A3Ventures*

ISSN 2640-3536

Introduction

State of the Industry

As set forth in the initial SAE EDGE™ Research Report: "Unsettled Topics Concerning Sensors for Automated Road Vehicles," the potential for automated driving systems (ADS) to reduce traffic congestion, accidents, transit times, and fuel consumption - while enabling personal mobility and new business opportunities - makes automation arguably the most important topic for the future of ground vehicles, if not mobility as a whole [1].

However, the development of these complex systems of systems requires substantial testing of their sensing, perception, and decision-making algorithms before they are deployed. Depending on a given operational design domain (ODD) - a specified scenario in which a self-driving vehicle can safely operate on its own - ADS may need to be tested and validated for a variety of different scenarios. Budget constraints and development times limit the amount of public-road and closed test-track driving that an ADS developer can feasibly conduct.

For ADS developers, this means a balanced mix of simulation in virtual environments, driving on closed test track courses, and driving on public streets and highways. These simulated environments - which can accumulate thousands of hours of simultaneous test driving - are a staple of the iterative and continuous life-cycle evolution of software development.

However, to achieve public trust and discover unexpected situations (or edge cases), vehicles must be tested on public roads - and before testing on public roads, ADS need to be trialed in physical, real-world test environments. Determining when and how these systems are tested remains a challenge for ADS developers and is the impetus for this SAE EDGE™ Research Report.

Unsettled Issues in Balancing Virtual, Closed-Course, and Public-Road Testing of Automated Driving Systems

In general, three techniques - virtual simulation, closed-course test tracks, and public-road driving - present a varied and comprehensive testing framework that can offer differing levels of environmental control and fidelity. Beyond testing in a single method, these techniques can be used in a layered, iterative, and complementary way to progressively evaluate an ADS [2].

However, there are some unsettled issues regarding how to balance ADS testing. The three most critical questions as identified by the experts contributing to this report include

- What determines how to test an ADS?
- What is the current, optimal, and realistic balance of simulation testing and real-world testing?
- How can data be shared in the industry to encourage and optimize ADS development?

Determining What Kind of Testing an ADS Requires

Modeling and simulation of an ADS, its subsystems, and its environment (including other traffic agents) in a virtual ecosystem can establish a baseline of how the ADS will act in the real world. This process requires appropriate modeling fidelity of all simulation components, including sensor, vehicle, environment, and driver models of both the ADS and all other traffic agents - including all vehicles, pedestrians, motorbikes, animals, etc. in the simulation (this is discussed in further detail in the SAE EDGE™ Research Report on "Unsettled Topics Concerning Sensors for Automated Road Vehicles" and "Unsettled Issues in Determining Appropriate Modeling Fidelity for Automated Driving Systems Simulation"). ADS behavior established through simulations is based on the fidelity of the simulation and its models: the higher the fidelity the more accurate the simulated behavior.

Pros and Cons of Different Types of Testing

Simulations also provide excellent agility, controllability, predictability, repeatability, and efficiency - serving as a cost-effective and systematic starting point for ADS development and testing. Simulations do not require the significant upfront costs of full vehicle development and test track facility resources. Additionally, ADS testing occurs at an accelerated pace with the increased scope of testing in a simulation environment and can include hazardous scenarios without endangering human life. Simulation tests also allow for determinism in the testing framework through repeatability (e.g., an ADS developer doesn't have to wait until sunrise every morning to test sun blinding for sensors). Alternatively, simulation does not uncover "never-before-thought-of" scenarios and is limited to the latest degree of achievable model fidelity. It is also important that simulations are mapped to real-world results.

Much like traditional driver-operated road vehicles, ADS should ideally be tested on a closed course or test track first to

FIGURE 1. **Notional ADS simulation architecture.**

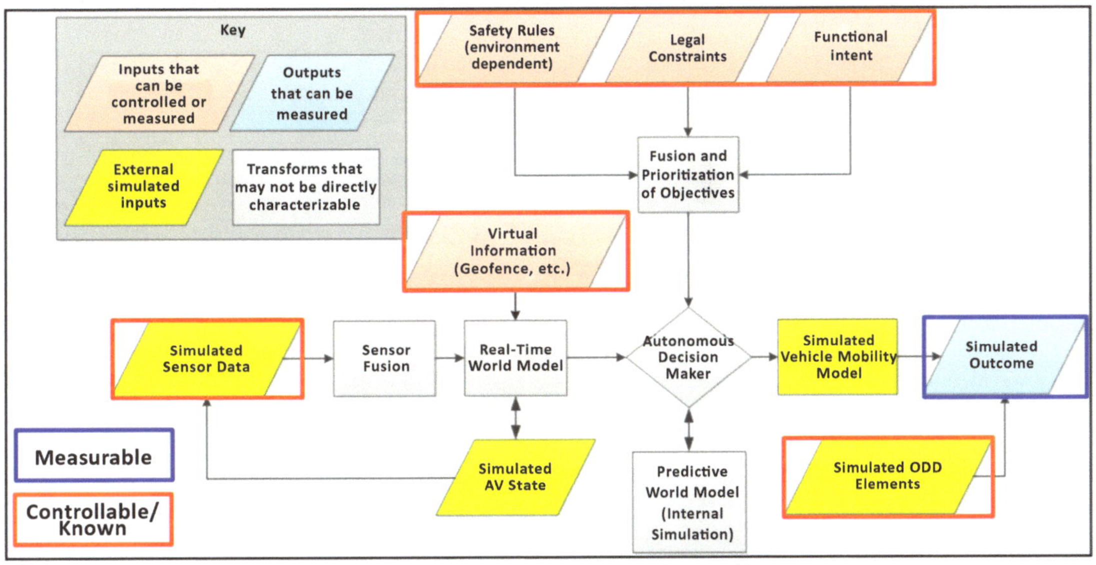

Reprinted from National Highway Traffic Safety Administration

establish vehicle handling behavior (and in the case of ADS - the validating baseline behavior established in simulation, and the simulation itself). These test tracks may be as simple as a large area of pavement or physical models of real-world environments, with parked cars, sidewalks, and roadside structures. They are still restricted to some degree based on their geometries, dimensions, size, and ODD limitations, but they do allow ADS developers to test scenarios that are deemed risky for public roads.

While test tracks won't uncover edge cases, they do provide a higher fidelity testing environment than simulation, without the danger of public-road tests. Many variables are controllable and test track scenarios can be transferred to other locations (like similarly plotted open-road scenarios) and/or virtual environments. New scenarios and cases uncovered by real-world driving that are modeled in a simulation help build more robust ADS testing - ideally, all cases and scenarios should be moved further and further up the test chain (to simulation environments). Additionally, while the scope of scenarios may be limited when compared to simulation (e.g., weather conditions, time of day, number and type of other traffic agents, etc.), test track scenarios are repeatable. Beyond limited scenarios, operating on test tracks is resource-intensive and introduces physical hazards and risk. Test tracks are also susceptible to the effects of weather, for better or worse.

Lastly, the shift toward ad hoc public-road testing allows for an extremely wide - if not uncontrollable - range of real-world variables and "unknown unknowns." While real-world, public-road testing doesn't provide repeatability or scalability, learned edge cases can be brought back to test tracks and/or simulation environments. Public-road driving also helps to build public trust in automated systems - the public typically has a limited understanding of simulation and test tracks.

Furthermore, tests on public roads can be conducted with the ADS in control of the vehicle or can be conducted in a "shadow mode" where a human test driver is operating the vehicle and the ADS system runs in the background but does not execute any control. This latter method can be used until the development team is comfortable enough with the decision-making process of the ADS (i.e., that it aligns close enough with a trained safety driver).

Figures 1-3 published in the National Highway Transportation Safety Administration's "A Framework for Automated Driving System Testable Cases and Scenarios" present the controllable inputs and measurable outcomes for the three main types of ADS testing [2].

Blended Testing

While these three test methods may be performed in sequence early in ADS development, they may be varied and mixed in order to inform and validate each other over the course of ADS development. They may even be layered into intermediate, blended forms of testing, such as simulating virtual events on a closed course with a real-life ADS. For example, a real ADS could test pedestrian avoidance on a real closed-course test track by using simulated pedestrian agents.

Blending virtual with real-world testing increases testing capabilities, but it also creates a spectrum of testing options. It may be difficult for ADS developers to determine what type of testing is needed. (Table 1 provides a snapshot of the strengths and weaknesses of the three primary test types.)

FIGURE 2. **Notional ADS track testing architecture.**

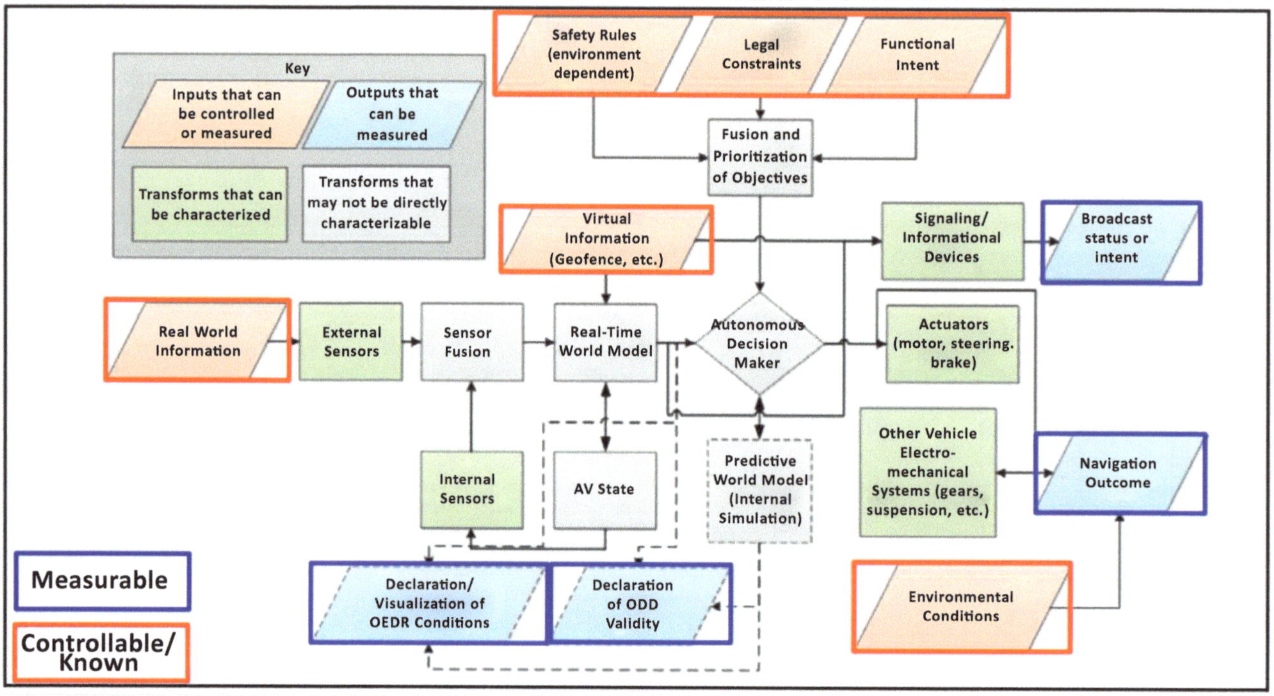

Reprinted from National Highway Traffic Safety Administration

FIGURE 3. **Notional ADS public-road testing architecture.**

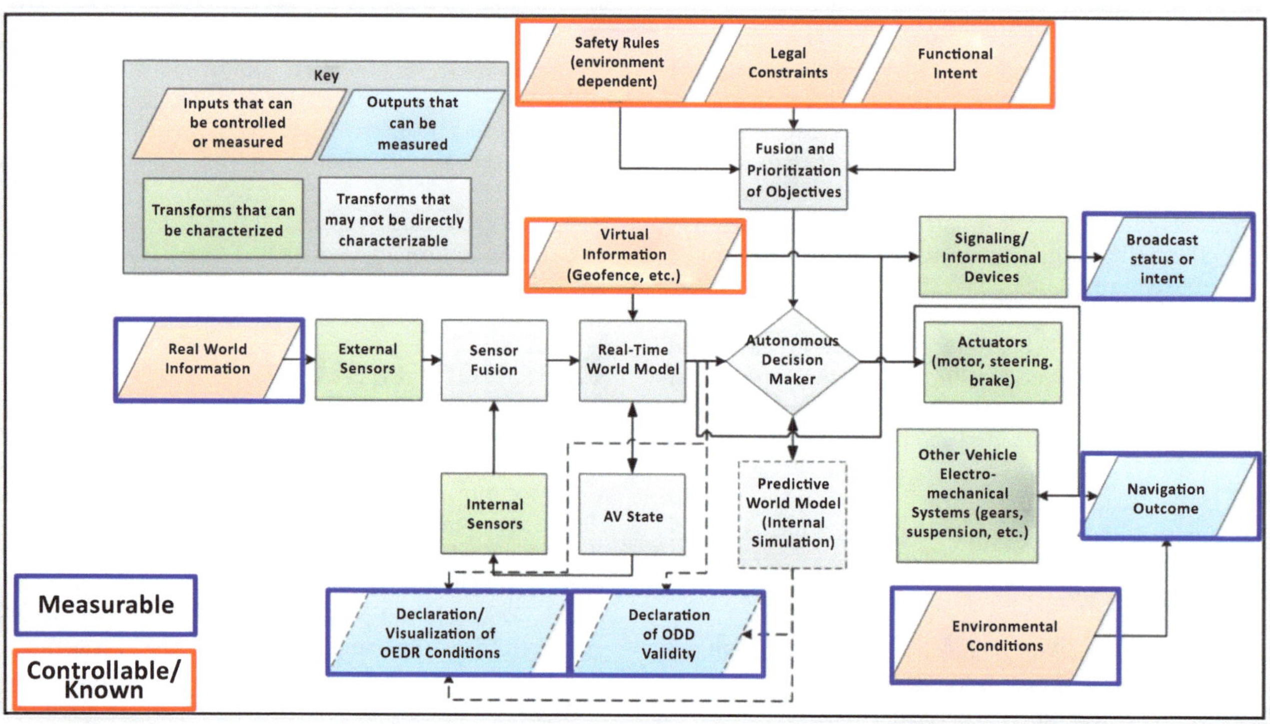

Reprinted from National Highway Traffic Safety Administration

TABLE 1. Pros and cons of ADS test types.

	Pros (+)	Cons (-)
Simulation	• Agile (systems changes can be reevaluated immediately) • Controllable (can be performed night and day) • Cost-effective and efficient (can be scaled and automated) • Predictable and safe (the scenario is controlled and no lives or physical hardware is at risk) • Systematic and repeatable (tests can be rerun, iterative changes can be applied, and previous versions can be recalled)	• Does not uncover edge cases and unknown unknowns • Limited by current state of modeling fidelity
Test track	• Safer than public roads while including physical hardware and actual vehicle dynamics • Very specific scenario and situation are as close as possible to real world • Risk can be anticipated and relatively easily accounted for	• Expensive • Limited by course infrastructure • There will always be small unanticipated variations
Public road	• Find out if system does what it is supposed to do • True operating environment • Uncovers unknown unknowns	• Can be hazardous • Expensive • Unrepeatable • Unpredictable

Ultimately, simulation, test track, and public-road testing aren't siloed environments, and blended testing is a clear example of this. While they may be discrete testing environments, they are very interrelated - and without looking at them in that way, ADS testing might be less effective and could miss safety-critical situations.

Cyclical Testing

Building from basic test environments (virtual, closed course, and open road) to blended test environments can provide unique and powerful means to assess an ADS. However, it is most important to note that testing is a cyclical process: real-world data must inform sensor data, which is then simulated, and then decision-making is tested until it is mature enough for public roads - this will continue to repeat and inform later cycles of testing.

Comparing Real and Virtual Miles

As mentioned before, the general public - the end users of ADS - as well as politicians (who affect federal, state, and local ADS testing legislation) tend to value public-road testing as a qualifier of system maturity, regardless of countless miles of testing conducted in a simulation environment. This holds true regardless of what type of public-road miles an ADS has accumulated during testing. For example, millions of miles of public-road testing on a highway may only present an ADS and developers with a narrow range of scenarios compared to tens of thousands of miles of driving in a high-traffic, urban environment. Yet public-road testing, which establishes public trust, also adds risk to public roads.

Several experts who contributed to this report mentioned that, in an ideal world, testing would completely rely on simulation. This approach is based on the notion that a simulation environment (given that it is modeled in high-enough fidelity) is the most efficient and safest way to test an ADS.

However, these real miles have the potential to uncover unanticipated ADS behaviors, and other experts contributing to the report hold that simulation will never remove the need for driving on public roads or test tracks. Even more forward-looking, some experts suggested that learning how automated vehicles affect normal road traffic (including pedestrian behavior) will be vital to the industry. For example, if an ADS is operating in ideal driving conditions (e.g., in a low-traffic area in a bright and clear environment) and sees a manhole cover which inadvertently causes it to brake, it may cause a rear-end accident if a human driver in a following car was not expecting the ADS to stop. Experts also mentioned examples of ADS not passing through an intersection with pedestrians standing at the corner, where neither the ADS nor the pedestrians moved due to uncertainty of whether the other party would move first. Standstills like this can cause dangerous scenarios for other actors in the traffic environment.

These scenarios that include third-party behavior in hybrid human-automated traffic are situations that ADS developers will ultimately want to test; however, this requires high-fidelity driver models and a documented understanding of human driver and pedestrian behavior [3]. Understanding automation's impact on human traffic and behavior is especially important when considering completely driverless shuttle vehicles like slow-moving buses (e.g., human drivers frustrated by automated vehicles may opt to pass them and potentially create hazardous traffic situations).

Determining What Type of Testing Is Appropriate

Some of the considerations between one test mode and another have been discussed to a degree in the previous section: time, economics, and risk are all determining factors for whether to test in real world or in a simulation. An ADS cannot be validated by one type of test. By balancing testing across different environments, scale, scope, and development times are weighted more equally. The enormous scope of an ADS ODD would be too much to accomplish at a proving ground, but avoidance maneuvers - for example - need to be tested in real world to validate previous simulations.

Determining the proper balance of testing leads to several questions, such as

- What sort of scenarios should developers absolutely test on public roads?
- What sort of scenarios should developers not test on public roads?
- Should developers only collect data on public roads?
- Should developers validate scenarios or only simulation models?
- Should a validated model be used in any scenario or does needs to be validated in each scenario?

The following sections provide insight from experts in ADS development who were asked how they would answer these questions.

What Type of Balance Is Needed?

Ideally, ADS developers should perform as much simulation as they can afford - many unpredictable scenarios occur in the real world, but in simulation one can accurately repeat the same case, especially previously established edge cases. Some of the experts involved in this report suggest that "more testing is better," and they could not imagine there is "too much" of any kind of testing.

Experts contributing to this report agreed that some ADS maneuvers (e.g., unprotected left-hand turns) are challenging to test, because they are very difficult to simulate with an appropriate level of real-world applicability. These types of scenarios therefore call for testing on the test track first and then on public roads.

Alternatively, certain hazardous situations may require the use of simulation and test tracks before a vehicle is deployed on public roads. For example, pedestrians stepping into the road might call for an iterative simulated, blended, and test track approach that involved a highly trained safety driver.

In contrast, public-road testing allows for requirements gathering for an ODD, which can cyclically inform both simulation and test track scenarios. Since the ODD is not constant (behaviors/infrastructure changes), continuous learning needs to occur. Additionally, public-road tests are needed to develop requirements because they reflect the ODD holistically. For example, a simulation may lack the randomness of a traffic situation; however, a simulation can very purely present an ODD without the challenge of unwanted embellishment (e.g., weather conditions like persistent rain when testing needs to occur in dry conditions).

What Minimum Requirements Should Be Met Before Public-Road Testing

While all ADS will go through simulation, closed-course, and public-road testing, it may be difficult to determine when a system is ready for testing on public roadways. This determination involves the ODD of a given ADS. For example, with automated shuttle vehicles (often called "robotaxis") operating in urban environments at lower speeds, there may be a lower threshold for shifting to public-road testing. Robotaxis are limited in that they are operationally constrained to pick-up and drop-off scenarios and are typically geofenced. On the other hand, an ADS destined for personal ownership will operate in a variety of other scenarios (e.g., high-speed highway driving) in addition to slower, urban navigation.

The threshold for whether an ADS is ready for public-road testing cannot be based simply on the number of miles or hours completed in simulation or closed-course testing. Experts contributing to this report suggested that an ADS needs to be tested in an increasingly complex and erratic series of scenarios before operating in public. These scenarios will likely be dynamic and relative to the respective development state of a respective ADS - that is, scenarios used at one point in time may not be adequate to test systems developed months later due to increased intelligence and competency (e.g., future ADS reaction times will decrease).

These thresholds may also be based on vehicle specifications such as horsepower, braking power, etc. A more capable vehicle may require more extensive testing because of the ability of engaging in highly dynamic driving that increases the scope of its ODD.

Furthermore, when an ADS is being tested in a public environment, it is being validated in its ODD. It's ODD cannot be fully replicated (yet) in total fidelity in simulation or a test track with the full extent of random factors that may be introduced. The question then shifts to "What level of exposure in that erratic environment is enough to qualify that an ADS can drive as well as or better than a human driver?"

Testing Software Changes on Existing Hardware

Once automated vehicles are on the road, there will be points at which their operating software needs to be fixed and/or expanded, and software updates will need their own balanced approach to testing.

If developers applied to each software update traditional automotive industry testing methodology like for instance the V-model, which foresees a large part of real-world testing, cost and time requirements would be very difficult to meet (Figure 4). The only way to scale is through simulation.

In software development, a set of regression tests would traditionally be performed that would qualify software changes. The same entire set of tests would be run, depending on whether it's a major or minor update to the software and then a developer would qualify it, and the decision would be made to deploy the update. However, what level should be tested at the proving ground or public roads?

For instance, if a change needs to be validated initially in general terms before a major system redesign might build on it (e.g., an additional system input which is initially only to be validated for general compatibility purposes, but ultimately going to interact with the system as a whole), ADS developers may be able to test it in simulation, but they still may require developers to create a new simulation model or scenario for the change. The question then becomes if a new ADS simulation models needs to be validated on a proving ground because the change might ultimately be so complex that it needs to be revalidated before taken on public roads.

The commercial aerospace industry serves as a good example for the ADS industry for certification of vehicle simulation and modeling.

Recommendations

Some of the experts involved with the development of this report suggest a Turing test could be used for L4 and L5 ADS systems (as defined by SAE J3016™), in particular to gauge the capability of their artificial intelligence (AI) and determine if a system is mature enough to move through more hazardous forms of testing. If the ADS driver model for the development vehicle operates as good as or better than a human controlling a vehicle in a virtual scenario, then the ADS could move on to a test track with partially simulated surroundings. The ADS can then be tested for how the system reacts after the introduction of a partially real, partially simulated vehicle agent. Then the next step would include operation in a highly controlled test-track environment, continuing by observing what the system perceives, processes, and executes. By observing the differences, ADS developers may be able to begin describing ADS capabilities and what stage the system is at in terms of its reactions and behaviors. To be clear, those decisions of moving on to the next step largely depend on the determination that a system or model "operates as good as or better than a human driver," as pointed out above. However, when this is accomplished, it will have solved one of the most-debated questions in the industry. This makes the determination when to move from

FIGURE 4. A V-model for project development [4].

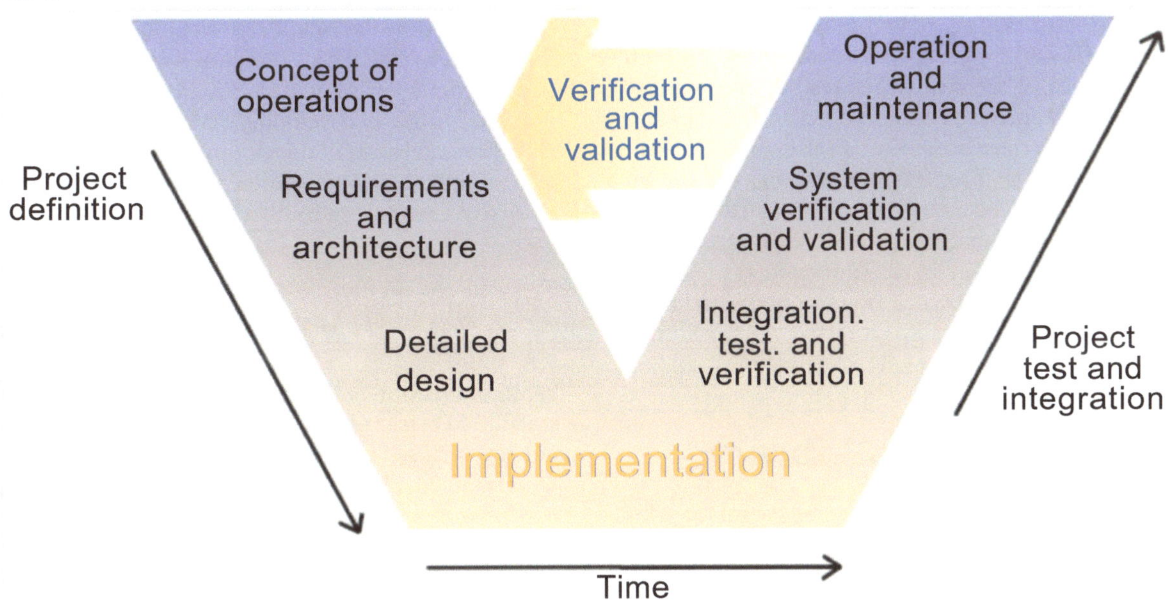

Reprinted from Clarus Concept of Operations. Publication No. FHWA-JPO-05-072, Federal Highway Administration (FHWA), 2005

one testing phase to the next very difficult, and ultimately reliant on expert's judgment.

This requires that ADS developers and testers enhance their human-driver training programs. Thorough safety-driver and human-driver training is needed to ensure that closed-course testing would be effective enough to result in an ADS that is ready to move forward into public-road testing.

The use of probabilistic/nondeterministic testing in simulation to identify unexpected situations is critical to ADS testing. This is similar to the development of AI for computers engineered to master games like chess or Go, where the computer is trained to play the game by losing less and winning more over time.

Professional judgment is key in the decision when to move from one stage to the next, and some of the experts involved with this report suggest that the nascent UL 4600: Standard for Safety for the Evaluation of Autonomous Products points to this [5]. A team that includes more than ADS developers should make the decision to determine system readiness. This could include lawyers and insurance experts.

Where Would Cross-Company Collaboration Be Advantageous?

Scenario-Based Data Sharing between Industry Players

While there is currently little data sharing between various organizations in the ADS industry, Germany's Pegasus Research Project - which aims to establishing criteria, tools, methods, and test scenarios for ADS - has suggested testing standardization and the "creation of a cross-manufacturer method for the safeguarding of highly automated driving functions." [6]. The working group has also called for common data format for sharing scenario-based testing information.

This could potentially enable an ADS developer to re-simulate parameterized scenarios identified by industry partners to identify ADS properties by whether the system passes a scenario, identified by industry as crucial to the validation of such systems. This could provide a workaround for intellectual property issues, giving ADS developers the ability to test and compare their ADS solutions without disclosing how the development was achieved internally (i.e., a simple pass/fail test).

Common Materials Library

A common materials database could also be useful for ADS development, as the differences between models and the real-world environments can be significant. Field testing often reveals whether environments were modeled with a high-enough fidelity [3]. For example, the appropriate modeling of asphalt plays a significant role in ADS simulation and testing. Information like friction and reflective properties of driving surfaces like asphalt (especially considering different weather conditions) are important to simulation. Additionally, differences in material properties such as between European and North American asphalt could have an impact on simulation which, in turn, informs ADS development.

Recommendations

Sharing scenario-based data can also increase awareness of edge cases uncovered through closed-track and public-road testing. There are certain scenarios that ADS developers do not test for yet, but the open access to a database of edge cases would mean that developers do not have to spend a vast number of expensive hours on public roads in search of unknown unknowns. While these cases (in their universal format) may be simple models, they can be used as a starting point for more sophisticated testing (e.g., conducting a simulation or real-world test with dirt or a scratch on a sensor lens). The progress towards shared scenario-based test data that could eventually lead to a test structure that heavily focuses on simulation is supported by closed track, and only in rare exceptions requires public-road testing due to the amount of established test cases.

These sharable scenarios could inform ADS developers and testers about ideal public-road test locations, such as locations that have a high concentration of special situations, so that hours spent conducting public-road testing have as much exposure to risk and chaotic conditions as possible. This could potentially go further to develop a framework for quantifying "smart miles," or miles that actually served to test an ADS thoroughly (as opposed to millions of miles of highway driving with no heavy stop-and-go traffic).

A catalog of materials such as road markings and retroreflective surfaces would be a valuable asset for ADS developers conducting simulations. Partnerships and/or sponsorships with respective materials companies could result in beneficial buy-in from relevant materials manufacturers and help alleviate the burden of developing a common materials database.

Ultimately, ADS development is a competitive landscape, but technology is universal and sharing information is critical - not just in industry but also with authorities.

Summary/Conclusions

While simulation may be the key enabler of efficient, safe, and scalable ADS testing, it requires validation through real-world testing on closed courses and public roads. These test formats are not siloed: they can and should overlap and be layered as part of an iterative and cyclical, ongoing test process.

Public trust of automated vehicles requires seeing an ADS operate on a public road; however, public-road testing needs to be conducted with safety as the absolute priority by operating in challenging erratic environments to pressure test an ADS and potentially unveil further edge cases.

Ultimately, deploying an ADS requires buy-in from not only the developers, but from legal and insurance experts. The ADS industry should look to aerospace best practices for guidance regarding automated systems development, testing, and certification.

SAE EDGE™ Research Reports

SAE EDGE™ Research Reports, like the present report on "Unsettled Issues in Balancing Virtual, Closed-Course, and Public-Road Testing of Automated Driving Systems," are intended to push further out into still unsettled areas of technology of interest to the mobility industry. SAE launches these reports before attempting to form a joint working group, let alone a cooperative research program or a Standards Committee.

SAE EDGE™ reports are intended to be quick, concise overviews of major unsettled areas where vital new technologies are emerging. By definition, an unsettled area is characterized more by confusion and controversy than established order. Early practitioners must confront an absence of agreement. Their challenge is often not to seize the high ground but to find common ground. These scouting reports from the frontiers of investigation are intended merely to begin the process of sorting through critical issues, contributing to a better understanding of key problems, and providing helpful suggestions about possible next steps and avenues of investigation.

SAE EDGE™ Research Reports, therefore, are fundamentally distinct from the more formal working groups approach and far removed from the more mature cooperative research programs and standard's development process. They begin by convening an informal meeting of subject matter experts working in an unsettled area of technology from industry, government, and the research community. A moderated but collegial discussion ensues regarding their shared interest. This conversation attempts to identify critical issues hindering a clear understanding of important topics, preventing a common agreement on potential solutions, or slowing progress in key areas of research.

An SAE EDGE™ Research Report is then drafted and circulated for comment to meeting attendees and other interested parties. After review, the final result of this process is a report like the current examination of unsettled issues concerning the balanced testing of automated vehicles.

Next Steps for Determining Appropriate Modeling Fidelity for ADS Simulation

This publication should be considered only as a first step toward clarifying the issues in balancing virtual, closed-course, and public-road testing of ADS. The intention behind this and other SAE EDGE™ Research Reports is to start a dialogue among interested parties on important industry-wide topics that require further attention. The expectation is that these explorations of unsettled areas of technology will lead to the formation of working groups and, ultimately, committees that can address and resolve the issues they raise, producing a framework for developing a common vocabulary of definitions, best practices, protocols, and standards needed to support continued progress toward safer and more innovative automotive products.

The meetings and interviews that gave rise to this publication demonstrated a great willingness on the part of the industry - and many of its experts - to define balance of testing methods and eventually the standards needed to enable testing procedures to move ahead as quickly and efficiently as possible. SAE International has demonstrated its lead in this and closely related areas by the introduction of unsettled sensor topics in the SAE EDGE™ Research Report on "Unsettled Issues in Determining Appropriate Modeling Fidelity for Automated Driving Systems Simulation," the basic reference work on active safety system sensors (SAE J3088™), the definitions for automated driving levels (SAE J3016™), the first vehicle cybersecurity framework (SAE J3061®), and numerous standards that establish the building blocks of ADAS, electrification, and intravehicular and extravehicular communications.

This SAE EDGE™ Research Report on "Unsettled Issues in Balancing Virtual, Closed-Course, and Public-Road Testing of Automated Driving Systems" identifies the following key topics for further pursuit, both through continued informal discussions among industry practitioners and through more formal working groups:

- Creating a catalog of material models to facilitate high-fidelity simulation tests
- Applying lessons learned from the commercial aerospace industry to ADS development
- Developing a scenario-sharing format and database as well as a framework for identifying locations or characteristics of ideal real-world test environments
- Creating a decision-making framework to determine when an ADS should initially transition from simulation to closed-course and from closed-course to open-road testing

Recommendations

The overall recommendations of this SAE EDGE™ report can be summarized as follows:

1. Turing tests could be used for L4 and L5 ADS systems (as defined by SAE J3016™) to gauge the capability of their AI and determine if the system is mature enough to move through more hazardous forms of testing. If the ADS driver model for the

vehicle in development operates as good as or better than a human directing a vehicle in a virtual scenario, then the ADS could move on to a test track with partially simulated surroundings and test how the system reacts after the introduction of a partially real, i.e., partially simulated vehicle agent. (However, quantifying what "good as or better than a human driver" means is necessary.) Then the next step would include operation in a highly controlled test-track environment, continuing by observing what the system perceives, processes, and executes. By observing the differences, ADS developers may be able to begin describing ADS capabilities and what stage the system is at in terms of its reactions and behaviors.

2. To identify unexpected situations through probabilistic/nondeterministic testing in simulation is critical to ADS testing. By applying probabilistic methods or AI, an ADS could be trained to drive similar to how other computers are trained to master games like chess or Go, where the computer repeatedly plays until it keeps winning more over time.
3. The emerging standard "UL 4600: Standard for Safety for the Evaluation of Autonomous Products" should be considered for insight into determining ADS maturity and readiness.
4. Scenario-based data can and should be shared to increase awareness of edge cases uncovered through closed-track and public-road testing. Open access to a database of edge cases would mean that developers do not have to spend a vast number of expensive hours on public roads in search of unknown unknowns.
5. A catalog of materials such as road markings and retroreflective surfaces would be a valuable asset for ADS developers conducting simulations. Partnerships and/or sponsorships with respective materials companies could result in beneficial buy-in from relevant materials manufacturers and help alleviate the burden of developing a common materials database.

Abbreviations/Definitions

ADS - Automated Driving System. Intelligent computer control of vehicles based on sensor input, characteristic of vehicles at L3 and higher *Automated Driving Levels*. See the entry for this term below.

Automated Driving Levels (as defined by SAE J3016™)

L0 - No Driving Automation. A human driver performs the entire Dynamic Driving Task (DDT). Other vehicle systems may provide warnings or support, e.g., braking systems with automatic skid control.

L1 - Driver Assistance. The driving automation system is capable of executing either the longitudinal or the lateral vehicle motion control subtask (not both). It disengages immediately upon driver request. The human driver is present at all times to perform all control tasks not done by the automation system, to determine when it is appropriate to engage the automation system, to supervise all operations of the automation system, and to intervene as necessary to maintain safe vehicle operation. Example: automatic speed control ("cruise control") for highway driving.

L2 - Partial Driving Automation. The driving automation system is capable of executing both the lateral and the longitudinal vehicle motion control subtasks. It disengages immediately upon driver request. A human driver must be present and attentive at all times to perform control tasks not done by the automation system, to determine when it is appropriate to engage the automation system, to supervise all operations of the automation system, and to intervene as necessary to maintain safe vehicle operation. Total driver disengagement from human control mechanisms (hands off the wheel and feet off the pedals) is possible in narrowly limited driving scenarios, e.g., lane centering.

L3 - Conditional Driving Automation. A human driver must be present and attentive at all times to operate the vehicle outside the ODD of the ADS and to take over control of the vehicle from the ADS within its ODD as necessary. The ADS can only be engaged in its ODD but can perform the entire DDT when engaged. It automatically disengages when it determines that either its ODD limits are about to be exceeded or it is experiencing a performance-relevant system failure, first, by issuing a timely request to intervene to the human driver and, second, by disengaging an appropriate time after issuing the request. It also disengages immediately upon driver request. The human driver is responsible for verifying the operational readiness of the vehicle, handling all DDT outside the ODD of the ADS, determining when the ADS can be engaged, and serving as the fallback-ready driver when the ADS is engaged, responsible for maintaining a minimal risk condition by actively intervening in ADS operation or assuming vehicle control on automatic ADS disengagement. An ADS ODD might include, for example, freeway driving or light traffic on public roads outside urban areas.

L4 - High Driving Automation. The human driver is responsible for verifying the operational readiness of the vehicle and determining whether to engage the ADS. The ADS will permit engagement only within its ODD. When the ADS is engaged, the human driver has no fallback role and may ride as a passenger. Once engaged, the ADS will transition to a minimal risk condition before it disengages (1) as the result of a performance-relevant system failure, (2) after an automatically issued request to intervene is ignored by a human driver, or (3) in response to a user request that it achieved a minimal risk condition. Optionally, a human driver may reassume DDT by requesting the ADS to disengage and waiting for it to turn over control of the vehicle once the ADS determines that conditions are appropriate or, in a fallback role, following an ADS request to intervene; or, after the ADS has automatically disengaged, following achievement of a minimal risk condition.

L5 - Full Driving Automation. The ADS ODD expands to permit engagement under all driver-manageable on-road conditions. L5 vehicle operation, when the ADS is engaged, is identical to L4 vehicle operation.

ODD - Operational Design (Driving) Domain, a specified scenario in which a self-driving vehicle can safely operate on its own, e.g., highway driving.

Acknowledgments

Recognition should go first to all the participants, many of who also provided feedback on the draft version of this publication. Without their input and initiative, this SAE EDGE™ Research Report would not have been possible.

Atul Acharya, AAA/A3Ventures
Michał Antkiewicz, University of Waterloo
David Barry, Multek Technologies
Thomas Bock, Samsung
Krzysztof Czarnecki, University of Waterloo
Eric Paul Dennis, PE, Center for Automotive Research
Cameron Gieda, AutonomouStuff
Peter-Nicholas Gronerth, fka Silicon Valley
Qiang Hong, PhD, Center for Automotive Research
Stefan Merkl, TÜV SÜD America
Chad Partridge, Metamoto
Robert Seidl, Motus Ventures
John Suh, Hyundai CRADLE
John Tintinalli, SAE International
Paul Wells, AAA/A3Ventures

Additional thanks go to the following individuals who also had the opportunity to provide feedback on the draft version of this publication. Their contributions helped shape this publication.

Gagan Aujla, Antenna Group
Gräfe Gunnar, 3D Mapping Solutions GmbH

The authors of this document together with the SAE Team responsible for its creation join in expressing our deepest appreciation to all the individuals mentioned above.

Sven Beiker, PhD
Silicon Valley Mobility
(external consultant to SAE International)

Works Cited

1. Beiker, S., "Unsettled Topics Concerning Sensors for Automated Road Vehicles," SAE International, 2019, https://saemobilus.sae.org/content/EPR2018001/.
2. Thorn, E., Kimmel, S., and Chaka, M., "A Framework for Automated Driving System Testable Cases and Scenarios," DOT HS 812 623, National Highway Traffic Safety Administration, https://www.nhtsa.gov/sites/nhtsa.dot.gov/files/documents/13882-automateddrivingsystems_092618_v1a_tag.pdf.
3. Beiker, S., "Unsettled Issues in Determining Appropriate Modeling Fidelity for Automated Driving Systems Simulation," SAE International, 2019, https://saemobilus.sae.org/content/EPR2019007/.
4. Santos, M.M., Mendes, C., Banik, T., Franco, F., and Neme, J., "New Approach of Tools Application for Systems Engineering in Automotive Software Development," SAE International, 2017, https://saemobilus.sae.org/content/2017-01-1601/.
5. Koopman, P., Ferrell, U., Fratrik, F., and Wagner, M., "A Safety Standard Approach for Fully Autonomous Vehicles," Edge Case Research and The MITRE Corporation, 2019, https://users.ece.cmu.edu/~koopman/pubs/Koopman19_WAISE_UL4600.pdf.
6. PEGASUS, "PEGASUS Research Project," https://www.pegasusprojekt.de/en/about-PEGASUS accessed November 21, 2019.

Contact Information

EDGEresearch@sae.org.

The authors of this document together with the SAE Team responsible for its creation join in expressing our deepest appreciation to all of the individuals who contributed.

www.ingramcontent.com/pod-product-compliance
Lightning Source LLC
LaVergne TN
LVHW070942210926
848482LV00003B/283
9781468601251